HOW TO MAKE AND SELL CABLES FOR BEGINNERS

Step-By-Step Guide To Creating High-Quality Sourcing Materials And Mastering Effective Techniques

REMINGTON BRIGGS

Copyright © [2024] by [Remington Briggs].

All rights reserved. Except for brief quotations included in critical reviews and certain other noncommercial uses allowed by copyright law, no part of this publication may be reproduced, distributed, or transmitted in any form or by any means, including photocopying, recording, or other electronic or mechanical methods, without the publisher's prior written permission.

DISCLAIMER

Greetings from the world of crafts! Before exploring the fascinating realm of creativity and craftsmanship that this book presents, we would like to make sure that the content is understood and can be clearly understood. This book's contents are meant solely for informational purposes. Although every attempt has been taken to guarantee accuracy and dependability, the material provided should not be used in place of or as a substitute for expert advice. Since every person's path with creating is different, we advise readers to use good judgment and, if necessary, seek out qualified professional guidance.

Crafting calls for imagination, trial & error, and individual interpretation. As a result, depending on personal abilities, tools utilized, and methods employed, outcomes may differ. The concepts, methods, or recommendations offered in this book are not guaranteed to produce any particular results, nor do the writers and publishers of the book. Additionally, it's critical to take safety precautions when working on crafts. Prioritize your own safety and well-being, use the proper tools and equipment, and always adhere to the manufacturer's recommendations.

We urge readers to be aware of their strengths and weaknesses and to experiment, be creative, and enjoy the making process.

TABLE OF CONTENTS

CHAPTER ONE 13
   CABLES OVERVIEW 13
      ELECTRICAL CABLE TYPES 13
      NETWORKING CABLE TYPES 15
      THE APPLICATIONS OF SPECIALTY CABLES 17
      EXAMINING AND CONTRASTING CABLE SPECIFICATIONS 19
      CHOOSING THE CORRECT CABLE FOR YOUR 21

CHAPTER TWO 23
   SUPPLIES AND EQUIPMENT 23
      ESSENTIAL COMPONENTS FOR MANUFACTURING CABLES 23
      KINDS OF INSULATORS AND CONDUCTORS 24
      INSTRUMENTS FOR RIPPING AND CUTTING WIRES 25
      EQUIPMENT FOR CRIMPING AND SOLDERING 27
      SAFETY EQUIPMENT AND MEASURES 28

CHAPTER THREE 29
   PROCEDURE FOR MANUFACTURING CABLES 29
      A COMPREHENSIVE GUIDE FOR CABLE ASSEMBLY 29
      METHODS OF SHIELDING AND INSULATING 31
      TECHNIQUES FOR BRAIDING AND TWISTING 34
      MEASURES OF QUALITY CONTROL 36
      TROUBLESHOOTING TYPICAL PROBLEMS 38

CHAPTER FOUR 41
   PACKAGING AND BRANDING 41
      ESTABLISHING A PERSONALITY FOR YOUR BRAND 41

- CREATING PACKAGING AND LABEL DESIGNS 43
- SUSTAINABLE PACKAGING SOLUTIONS 45
- RESPECT FOR LABELLING REQUIREMENTS 47
- EFFECTIVE PRODUCT MARKETING 48

CHAPTER FIVE 51
- ANALYSIS OF COSTS AND PRICES 51
  - COMPREHENDING PRODUCTION COSTS 51
  - ESTABLISHING COMPETITIVE RATES 52
  - PRICING TECHNIQUES FOR VARIOUS MARKETS 53
  - HOW TO DETERMINE PROFIT MARGINS 54
  - PROVIDING SALES AND MARKETING 55

CHAPTER SIX 57
- DISTRIBUTION AND SALES CHANNELS 57
  - INTERNET SALES: CHANNELS AND TECHNIQUES 57
  - OPPORTUNITIES FOR RETAIL AND WHOLESALE 59
  - PARTICIPATING AT EXPOS AND TRADE SHOWS 61
  - CREATING ALLIANCES WITH DISTRIBUTORS 63
  - ORGANIZING STOCK AND SHIPPING 65

CHAPTER SEVEN 69
- PROMOTION AND MARKETING 69
  - FORMULATING A MARKETING STRATEGY 69
  - MAKING USE OF INTERNET AND SOCIAL MEDIA ADS 70
  - BLOGGING AND CONTENT MARKETING 72
  - NEWSLETTERS AND EMAIL CAMPAIGNS 73
  - DEVELOPING CONNECTIONS AND NETWORKING 75

CHAPTER EIGHT ... 77
CUSTOMER ASSISTANCE AND SUPPORT ... 77
- OFFERING TOP-NOTCH CUSTOMER SUPPORT ... 77
- REFUNDS AND COMPLAINTS ARE HANDLED ... 79
- CONSTRUCTING A LOYALTY SCHEME ... 80
- OBTAINING AND EXAMINING CLIENT INPUT ... 82
- ENHANCING PRODUCTS IN RESPONSE TO CUSTOMER INPUT ... 84

CHAPTER NINE ... 87
GROWING YOUR COMPANY ... 87
- GROWING YOUR PRODUCT OFFERING ... 87
- EMPLOYING AND EDUCATING PERSONNEL ... 89
- CONTRACTING OUT PRODUCTION ... 91
- PURCHASING NEW TECHNOLOGY ... 92
- PLANNING AND MANAGEMENT OF FINANCES ... 94

CHAPTER TEN ... 97
FAQS & FREQUENTLY ASKED QUESTIONS ... 97
- SOLVING ISSUES WITH MANUFACTURING ... 97
- HANDLING COMPETITION IN THE MARKET ... 98
- TAKING CARE OF LEGAL AND COMPLIANCE ... 99
- RESPONDING TO INQUIRIES FROM CUSTOMERS ... 100
- GETTING READY FOR UPCOMING INNOVATIONS AND ... 101

ABOUT THE BOOK

Whether you're a hobbyist or an aspiring entrepreneur, the book "How to Make and Sell Cables" is a vital reference for anyone wishing to get into the cable manufacturing business. The introduction establishes the scene by examining the wide universe of cables, dissecting the various kinds of networking and electrical cables, as well as specialty cables and their applications. Producing high-quality products that satisfy consumer requests requires an understanding of the numerous specifications and knowing how to choose the appropriate cable for a given set of requirements.

The book then explores the necessary supplies and equipment needed to make cables. This section offers a thorough overview of the physical parts and tools required, including the different kinds of conductors and insulators as well as the instruments required for cutting, stripping, soldering, and crimping. To guarantee the efficiency and safety of the manufacturing process, safety equipment and measures are also stressed.

A detailed guide detailing the steps involved in cable manufacturing, including insulation, shielding, twisting, and braiding, follows the section on materials and tools. To ensure that readers can build cables that satisfy industry standards and customer expectations, quality control measures and typical issue troubleshooting are covered in detail.

Selling cables requires effective branding and packaging, and this book offers helpful advice on how to do these things. Effective marketing techniques and labeling standards compliance are also explored, assisting readers in showcasing their goods to prospective buyers in a polished and appealing manner.

Operating a profitable cable manufacturing company requires careful consideration of pricing and cost. The book provides a comprehensive overview of manufacturing costs, competitive pricing strategies, and profit margin calculations. It also goes over alternative pricing approaches for distinct markets as well as the advantages of discounts and promotions.

The examination of sales channels and distribution strategies provides readers with a thorough grasp of how to connect with their intended audience. This includes attending trade exhibitions and expos, investigating retail and wholesale prospects, establishing connections with distributors, and selling online via a variety of platforms. There are also effective shipping and inventory management strategies offered.

Business growth depends on marketing and promotion, and this book helps readers create a strong marketing strategy. There is talk of using social media, online advertisements, blogging, content marketing, email campaigns, and networking as ways to increase awareness and interaction with potential clients.

Sustaining a faithful customer base requires providing excellent customer service and assistance. The book places a strong emphasis on the value of offering top-notch customer service, managing returns and complaints, creating loyalty programs, obtaining and evaluating client feedback, and consistently enhancing products in light of this input.

The book provides helpful guidance on growing product lines, employing and training employees, outsourcing manufacturing, bringing in new technologies, and financial planning for people aiming to grow their firms. Gaining these insights is essential to growing a cable manufacturing company.

The book answers frequently asked questions (FAQs), solves manufacturing issues, offers tactics to counter rivalry in the market, handles legal and regulatory matters, responds to customer inquiries, and gets ready for new developments and trends in the cable sector. With the help of this thorough book, readers will be equipped with all the knowledge and resources necessary to thrive in the cutthroat world of cable sales and manufacturing.

CHAPTER ONE

CABLES OVERVIEW

ELECTRICAL CABLE TYPES

In many applications, electrical cables are necessary for transmitting electricity, and choosing the appropriate one requires a grasp of their varieties. The most popular kinds are coaxial cables, which are perfect for sending internet data and television signals since they have an outside cover, an insulating layer, a metallic shield, and a central conductor. Another kind is the twisted pair cable, which is made up of two insulated copper wires twisted together.

Because of its capacity to lessen electromagnetic interference, twisted pair cables are frequently utilized in data and phone networks.

Another important category is power cables, which are made to transport electrical power. They are available in different varieties, including flexible non-armored cables and armored cables, which have a metal covering for protection. Non-armored cables are frequently used in residential wiring, however, armored cables are usually found in industrial settings where durability is crucial. Safe and effective power transmission is ensured by the grading of each type of power cable for certain voltage levels and environmental circumstances.

In addition, there are ribbon cables, which are mostly used for internal connections in electronic devices and are made up of several conducting wires arranged parallel to one another. These cables are renowned for being flexible and having a small form factor, which makes them appropriate for usage in digital devices like computers. Making informed judgments for a variety of electrical applications requires knowledge of the many types of electrical cables, ensuring that the appropriate cable is utilized for the job.

NETWORKING CABLE TYPES

Networking cables are essential for creating and managing communication networks, and depending on the requirements of the network, there are several varieties to take into account. Cat5, Cat6, and Cat7 Ethernet cables are some of the most widely used types.

While Cat6 and Cat7 cables offer faster speeds and greater performance, making them ideal for modern, high-speed networks in both residential and commercial settings, Cat5 cables support rates up to 100 Mbps, making them adequate for basic home networking.

Another important category is represented by fiber optic connections, which use light to send data over great distances quickly. There are two primary varieties of these cables: single-mode and multi-mode. Multi-mode cables are used for shorter distances and are frequently found in local area networks (LANs) within buildings, whereas single-mode fiber optic cables are utilized for long-distance communication due to their capacity to retain signal integrity over large distances.

The last kind of Ethernet cable is called a crossover cable; it connects two network devices directly to each other without the use of a switch or router. These are very helpful when directly connecting PCs to share files or play games on a network.

Installing dependable and effective communication networks requires an understanding of the many kinds of networking cables and their uses.

THE APPLICATIONS OF SPECIALTY CABLES

Specialty cables are made for particular uses where certain features or functionalities are needed. To transfer high-definition audio and video signals between devices like televisions, Blu-ray players, and gaming consoles, for example, HDMI cables are frequently used in home entertainment systems. These cables are necessary for contemporary audio-visual setups because they provide clear, high-quality signal transmission.

Another kind of specialized cable is the USB cable, which is necessary for connecting and powering a variety of gadgets, including printers, external hard drives, and smartphones and tablets. There are several types of USB cables, such as micro-USB, USB-A, USB-B, and USB-C, each designed for a certain device or use. In both personal and business contexts, USB cables are essential due to their adaptability and simplicity.

There are cables made specifically to withstand challenging conditions and guarantee dependable operation in industrial and medical contexts. Silicone cables, for instance, are flexible and heat-resistant, making them appropriate for use in high-temperature environments or situations requiring frequent bending and movement. Similar to this, shielded cables guard against electromagnetic interference, guaranteeing the steady operation of machinery and delicate medical equipment.

You can choose the ideal cable for your particular needs if you are aware of specialty cables and their applications.

EXAMINING AND CONTRASTING CABLE SPECIFICATIONS

Comparing specs is essential when choosing cables to make sure they fulfill application requirements and performance standards. Conductor material, insulation type, and shielding are important parameters. Cable conductivity and flexibility are influenced by conductors, which are usually composed of copper or aluminum.

Because of its greater conductivity and longevity, copper is frequently chosen over aluminum, which is more cost-effective and lighter yet still useful in some situations.

Rubber, Teflon, and PVC are examples of insulating materials that safeguard the conductor and guarantee safety. Whereas rubber gives more flexibility and durability, making it ideal for cables that will be bent or moved frequently, PVC is extensively used since it is affordable and provides enough protection in most conditions. Teflon insulation is perfect for specific industrial applications because of its superior resilience to chemicals and high temperatures.

Another important requirement is shielding, especially for cables used in areas where electromagnetic interference (EMI) is prevalent. Stable signal transmission is ensured by shielded cables' metallic covering, which shields the conductor from outside electromagnetic interference. Unshielded cables work best in low-EMI conditions even though they are more flexible and affordable. You can select a cable that offers the best performance and dependability for your particular requirements by comparing these features.

CHOOSING THE CORRECT CABLE FOR YOUR REQUIREMENTS

It's important to comprehend the particular requirements of your application and match them with the suitable cable type and specifications when selecting the best cable for your purposes. Determine the purpose of the cable first, whether it is for data transfer, power transmission, or specialized uses like USB or HDMI connections. This first stage reduces the number of possibilities and concentrates on cables made specifically for your needs.

Next, think about the cable's exposure to various environmental factors, including temperature, humidity, and possible physical strain. For example, while basic PVC-insulated cables may suffice for household applications, cables used in industrial settings may require additional protection and durability, such as silicone or armored cables. Recognizing the requirements of the environment guarantees the lifetime and dependable operation of the cable.

Lastly, consider the technical details, like the type of insulation, conductor material, and shielding. Make that the cable satisfies your application's requirements for voltage, current, and data transmission rates. For instance, superior performance over Cat5 cables is guaranteed when utilizing Cat6 Ethernet cables for high-speed internet connections.

CHAPTER TWO

SUPPLIES AND EQUIPMENT

ESSENTIAL COMPONENTS FOR MANUFACTURING CABLES

You'll need the necessary materials that guarantee performance and durability to build high-quality cables. First things first, conductors are needed. These are usually copper wires used to carry electrical impulses. Because of its affordability and conductivity, copper is recommended. Equally important are insulators, which encircle the conductors to guarantee safety and stop electrical leakage. Teflon, PE (polyethylene), and PVC (polyvinyl chloride) are examples of common insulators.

To insulate the wire from radio frequency interference (RFI) and electromagnetic interference (EMI), you'll next require shielding materials. For this purpose, braided copper shields or aluminum foil are frequently utilized. You'll also need a jacket material, which encloses the entire cable assembly and offers insulation and mechanical protection. Depending on the intended usage of the cable and external conditions, jackets are frequently made of thermoplastic, rubber, or PVC materials.

KINDS OF INSULATORS AND CONDUCTORS

The selection of conductors and insulators in cable construction has a major influence on both performance and safety. There are several different kinds of conductors, such as stranded and solid conductors. Solid conductors are stable single-wire systems that are perfect for long-term installations. Stranded conductors are flexible and appropriate for applications that need to be bent frequently since they are composed of several smaller wires twisted together.

There are several varieties of insulators, each with special qualities. One popular insulator that is well-known for its adaptability and low cost is PVC. It can't resist extremely high or low temperatures, but it works well for general-purpose cables. Because of its weather resistance, PE (polyethylene) is another insulator that is utilized for outdoor cables. Teflon insulators provide exceptional high-temperature durability and thermal stability.

INSTRUMENTS FOR RIPPING AND CUTTING WIRES

Specialized instruments for cutting and stripping are required to operate with cables efficiently. For a clean and damage-free way to cut cables, cable cutters are needed. Wire strippers and cable stripping machines are examples of instruments used to remove insulation from conductors without damaging them. These instruments guarantee accuracy and effectiveness when setting up cables.

Crimping tools are required for making safe connections between wires and connectors in addition to cutters and strippers. The process of crimping creates a robust mechanical and electrical connection by compressing a metal sleeve, or crimp, onto the wire and connector. In cases where soldered connections are preferable over crimping, solder and soldering irons are also utilized to join wires.

EQUIPMENT FOR CRIMPING AND SOLDERING

In cable assembly, soldering and crimping are crucial skills. Solder wire, soldering stations, and soldering irons are examples of soldering equipment. Using heat, solder is melted and a bond is formed between conductors or components using soldering irons. Crimping tools and terminals are examples of crimping equipment. Without the necessity for soldering, crimping tools squeeze the terminal onto the conductor to create a dependable connection.

To prevent the solder from overheating or underheating while soldering, make sure the soldering iron is at the proper temperature. To increase solder adherence and flow, use flux. Choose the right crimping tool and terminals for the job by considering the type of connector and wire gauge.

To achieve strong and secure connections, crimp according to the manufacturer's instructions.

SAFETY EQUIPMENT AND MEASURES

Safety equipment and procedures are essential when working with cables because there are possible risks. When cutting and stripping, put on safety glasses to shield your eyes from flying debris. To avoid electrical shocks, wear insulated gloves when working with live wires. Maintain adequate ventilation in work environments, particularly during soldering, to prevent fume inhalation.

To avoid electrical mishaps, disconnect power sources if you are working on cables. To lessen trip hazards and cable damage, arrange and secure cables using cable management equipment like cable ties and clips.

CHAPTER THREE

PROCEDURE FOR MANUFACTURING CABLES

A COMPREHENSIVE GUIDE FOR CABLE ASSEMBLY

The first step in the cable assembly process is choosing the appropriate components, which are usually protective jackets, insulating materials, and conductors (which are commonly made of copper or aluminum). To start, precisely cut the conductors to the required length to prevent waste. After that, use a wire stripper to remove the conductor ends so that the metal is visible. This is an important step because it gets the wire ready to have terminals or connectors attached to it. Using crimping tools, gently attach the terminals or connectors after stripping. The performance and longevity of the cable depend on a secure connection, which is ensured by proper crimping.

Assembling the conductors into a cable comes next after they are ready. Depending on the type of cable, this entails arranging the conductors into the appropriate configuration, which may be twisted, bundled, or parallel. To keep the conductors organized in complicated cables, use cable ties or markers. Once arranged, place the conductors within protective tubing or an insulating jacket. This jacket offers the required defense from external elements and bodily harm. To keep the cable from moving and preserve its integrity, make sure the jacket is securely fitted around the conductors.

Test the cable's electrical performance and continuity to finish the assembly. To look for any breaks or flaws in the conductors, use a multimeter. To identify any mistakes early on, it is crucial to run these tests before finishing the assembly. After testing, make sure the connectors and terminals are correctly secured and tighten any loose ends. This comprehensive manual streamlines the process of assembling cables, guaranteeing a dependable and operational result suitable for a wide range of uses.

METHODS OF SHIELDING AND INSULATING

In the production of cables, insulation and shielding are essential elements that offer protection from outside interference as well as electrical insulation. Start the insulation process by choosing the right materials, such as rubber, Teflon, or PVC, based on the intended use of the cable and the surrounding conditions.

To guarantee safety and avoid electrical shorts, the insulating material is extruded entirely over the conductor. The insulating material is heated throughout the extrusion process until it becomes malleable, at which point the conductor is uniformly coated.

The next stage is to add shielding after the insulation has been installed. The cable is shielded against radio frequency interference (RFI) and electromagnetic interference (EMI). The two most common methods of shielding are braiding and foil wrapping or both combined.

Braiding offers superior flexibility and durability by encircling the insulated conductor with tiny metallic strands woven around it. In contrast, foil wrapping provides a lightweight and efficient screen against interference by encircling the conductor with a thin layer of copper or aluminum foil. The intended usage of the cable and the necessary level of interference protection determine which shielding technique is used.

To keep the shield effective after application, make sure it is correctly grounded. This entails attaching the shield to a grounding point, which is often found at one or both cable ends. An appropriate grounding system eliminates unwanted electrical noise, keeping the cable from functioning improperly. Check the shielding and insulation as well for any flaws or irregularities. This guarantees the cable will function dependably in the specified application, offering safety and interference-free protection.

TECHNIQUES FOR BRAIDING AND TWISTING

To improve mechanical strength and reduce electromagnetic interference, twisting and braiding are crucial methods in the production of cables. To create a twisted pair, two or more conductors must be wound around one another. To reduce crosstalk and interference, data and communication cables frequently employ this technique. To maximize performance, the twist rate—that is, the number of twists per unit length—is carefully regulated. The conductors should first be aligned in a twisting machine before being fed through the twisting mechanism. A uniform twisted pair is formed as a result of the machine's consistent twisting of the wires throughout the operation.

In contrast, braiding is the process of weaving together several strands of yarn or wire to create a jacket or shield. This method offers superior protection against mechanical stress and flexibility. The first step in braiding a cable is to load a braiding machine with the necessary number of spools, each of which holds a wire or yarn strand. After that, the device intertwines the strands to round the conductors in a taut, even braid. For added durability and protection, braiding can be used as an outer jacket or over-insulated conductors.

For best results, twisting and braiding need to be done with exact precision and consistency. Once the braiding or twisting is finished, it is crucial to check the cable for consistency and quality. The electrical and mechanical characteristics of the cable may be impacted by any anomalies. When cables are twisted and braided correctly, they reduce interference and increase strength, which makes them ideal for a variety of uses in industrial settings, telecommunications, and data transfer.

MEASURES OF QUALITY CONTROL

A vital component of the cable manufacturing process is quality control, which guarantees that the finished product satisfies all requirements. To make sure the raw materials—such as conductors, insulation, and shielding materials—meet the required quality standards, a thorough inspection is conducted at the start of the process. To avoid problems with the finished product, any defective materials are rejected. Continuous monitoring is crucial during the manufacturing process. To ensure that dimensions, such as conductor diameter, insulation thickness, and total cable diameter, are consistent, use automated inspection methods. To preserve quality, these systems can identify deviations and initiate corrections.

To verify the operation of the cable, conduct electrical testing in addition to dimensional checks. Testing for high-voltage breakdown, insulating resistance, and continuity are all included in this. Insulation resistance testing makes sure the insulation offers sufficient protection, while continuity tests confirm there are no breaches in the conductors. High-voltage tests verify the safety and dependability of the cable by examining its capacity to tolerate voltages greater than its rated level. These tests have to adhere to industry standards and are carried out with specialized equipment.

Lastly, conduct a thorough last examination before packing and sending the wires. This comprises functional tests to verify that the cable satisfies all performance requirements in addition to visual inspections for physical flaws such as cuts, abrasions, or misaligned components. To ensure traceability and for future use, record all inspection and test results. To ensure that the cables are dependable, safe, and fulfill the specifications needed for the purposes for which they are intended, quality control procedures are essential in the manufacturing of cables.

TROUBLESHOOTING TYPICAL PROBLEMS

Problems can still occur in the creation of cables even with meticulous manufacturing procedures. Conductor breakage, insulation flaws, and shielding defects are typical issues. Checking for physical damage or manufacturing flaws in the afflicted area is the first step in troubleshooting conductor breakage. To check for continuity and pinpoint the precise position of the break, use a multimeter. Re-strip and re-terminate the conductor after it has been located to guarantee a secure connection. To prevent such breakdowns, put preventive measures in place, such as appropriate handling and storage.

Deteriorated performance or electrical shorts are common signs of insulation defects. Examine the insulation visually for any cuts, abrasions, or uneven thickness to remedy these. To gauge how effective the insulation is, use an insulation resistance tester. Re-insulate the impacted areas using the proper materials and methods if flaws are found. Make sure the materials are of the highest caliber and that the insulation procedure is followed consistently. To avoid such problems, calibrate and maintain insulation equipment regularly.

The performance of the cable may be impacted by shielding flaws that enhance electromagnetic interference. Examine the shield for fractures, inadequate covering, or grounding concerns to troubleshoot shielding issues. To gauge how well the cable shields, use an EMI tester. Make sure the grounding connections are made properly and fix any cracks or holes in the shield. To guarantee constant shielding quality, test and maintain braiding and wrapping equipment regularly. Manufacturers may minimize performance concerns and continue to produce high-quality cables by quickly and effectively resolving these typical challenges.

CHAPTER FOUR

PACKAGING AND BRANDING

ESTABLISHING A PERSONALITY FOR YOUR BRAND

Establishing a distinctive personality and image for your cable products that appeal to your target market is part of building a brand identity. Establish your brand's principles, goals, and vision first. Think about the features that distinguish your cables from rivals. Which is it: eco-friendliness, durability, or creative design?

Create an engaging brand narrative that draws attention to your unique selling propositions and appeals to potential clients once you have a firm grasp of them. From your website and social media presence to your tagline and logo, all of your marketing materials should tell the same message.

Next, concentrate on the visual components that best capture your brand's essence. Select a color scheme, font, and design aesthetic that is in line with your brand's core principles and appeals to your target market. Take eco-friendliness, for example, into consideration when choosing colors and designs for your brand. Make a distinctive logo that people will remember and associate with your business. Make sure that every visual component—including packaging, ads, and internet presence—is utilized consistently and cohesively over all channels.

Establish the tone and voice of your brand lastly. This entails selecting the tone in which you speak with your clients, be it informal, kind, or hilarious. All written content, including product descriptions, blog entries, and customer care correspondence, should have a consistent brand voice. By upholding a consistent brand identity, you increase consumer awareness and trust, which enhances the attraction and memorability of your items.

CREATING PACKAGING AND LABEL DESIGNS

Creating compelling labels and packaging for your cables is essential to drawing in customers and getting the message across. Begin by choosing packaging materials that both showcase your brand identity and preserve your cables.

The packaging needs to be both aesthetically pleasing to draw in potential customers and robust enough to withstand harm during transit. Use bespoke packaging such as boxes, pouches, or wraps that feature your logo, colors, and other design components.

Make sure the labels are aesthetically pleasing, informative, and easy to read while developing them. Add pertinent details like the product name, specifications, how-to manual, certifications, and safety alerts. Make it easier for clients to find what they need by using an organized layout and a readable font. To make the label design blend nicely with the rest of your packaging, incorporate brand aspects like your logo and colors.

Consider how your labels and packaging might improve the customer experience as well. Packaging that is simple to open, alternatives that can be sealed, or sections for cable accessories can increase utility and convenience. Adding a QR code to the label that directs consumers to a product page or video lesson might further engage them and offer more information. You can make packaging that not only protects and informs but also improves the whole customer experience by putting equal emphasis on usefulness and beauty.

SUSTAINABLE PACKAGING SOLUTIONS

The importance of eco-friendly packaging is rising as more and more customers seek out sustainable and ecologically friendly goods. Choose materials that are recyclable, biodegradable, or composed of recycled resources first. Alternatives like paper, cardboard, and some kinds of bioplastics can be great options. Make sure the materials you select can survive the rigors of shipping and handling without compromising the protection of your cables.

Create your package with the least amount of trash and environmental impact possible.

This can entail creating packaging that has several uses, like a storage container that the buyer can reuse, or choosing smaller, more eco-friendly packaging. If you want to use designs that are simpler and use fewer resources, stay away from extraneous details and non-recyclable parts.

Make a statement about your dedication to sustainability with your labels and packaging. Give directions on how to properly dispose of or recycle the packaging, making it clear which portions are recyclable. Emphasize any environmentally responsible actions you do, like buying carbon offsets or utilizing plant-based inks. Making eco-friendly packaging a top priority will help you appeal to a rising market of environmentally concerned customers while also lessening your impact on the environment.

RESPECT FOR LABELLING REQUIREMENTS

Maintaining adherence to labeling guidelines is crucial to staying out of trouble with the law and earning your consumers' trust. Investigate the particular labeling regulations that apply to your product category and target market first. These specifications, which differ depending on the area, could contain details on electrical certifications, safety standards, and particular warnings. To prevent fines or recalls, make sure your labels contain all the required information.

Make sure that your labels prominently feature the necessary certifications and compliance markers. Symbols like the UL certification, RoHS compliance, or CE mark can be examples of this. These certifications offer clients peace of mind that your cables adhere to strict safety and quality standards. Make sure that these symbols are printed on the packaging in a noticeable and accessible manner.

To ensure compliance with any regulatory changes, examine and update your labels regularly. This may entail adding new certifications, revising the information's structure, or updating safety alerts. Keeping your labels current not only guarantees compliance but also shows your dedication to upholding strict guidelines and giving your clients trustworthy, safe items.

EFFECTIVE PRODUCT MARKETING

For your cable products to be successful, you must use effective marketing. Determine who your target audience is and get familiar with their requirements and preferences first. Create a marketing plan that draws attention to the special qualities and advantages of your cables, such as their exceptional durability, creative design, or environmentally responsible construction. Make sure your messaging speaks to your target audience's specific pain areas and resonates with them.

Make use of multiple marketing channels to expand your audience. This can involve using email marketing, social media, and search engine optimization (SEO) to increase website traffic. Provide interesting information that highlights the uses and advantages of your cables, such as blog articles, tutorials, and movies. To increase your reach and reputation, work with industry professionals or influencers to review and promote your items.

Lastly, evaluate the success of your marketing initiatives and make any necessary corrections. To evaluate the success of your efforts,
monitor important indicators including website traffic, conversion rates, and customer reviews. Utilize this information to hone your tactics, concentrating on the platforms and messaging that yield the greatest outcomes. You can boost sales of your cable products, draw in additional clients, and raise awareness by consistently refining your marketing strategies.

CHAPTER FIVE

ANALYSIS OF COSTS AND PRICES

COMPREHENDING PRODUCTION COSTS

You must first dissect the parts to comprehend the production expenses for cables. This covers connections, packing, insulating materials, and raw materials like copper or fiber optic strands. You can determine the cost of each component by obtaining estimates from suppliers or looking up industry statistics. Don't forget to account for labor expenses, which include paying technicians and assembly line personnel for quality control inspections.

Next, take into account overhead expenditures including the rent for the production facility, utilities (such as water and electricity), equipment upkeep, and administrative charges (such as management and support staff wages). These expenses are not directly related to specific cable units, but they are necessary for the production process. You'll be able to determine the exact cost of production for each cable unit by comprehending and computing these expenses.

ESTABLISHING COMPETITIVE RATES

Understanding the willingness to pay off your target audience and conducting a market analysis is necessary for setting competitive prices for your cables. To determine their price ranges, start by looking at comparable cables available in the market. When comparing pricing, take into account aspects like customer service, features, quality, and company reputation.

Once you have a firm understanding of market prices, identify the USPs that set your cables apart from those of your competitors. This could include outstanding customer service, cutting-edge features, or better quality. Make a case for pricing that corresponds to the value that clients would obtain by using these USPs. Remember that having a competitive price doesn't always equate to being the cheapest; rather, it means providing value that supports your asking price.

PRICING TECHNIQUES FOR VARIOUS MARKETS

Take into account variables like market demand, client demographics, and geographic location when putting pricing plans into practice for various markets. For instance, you may modify rates by the cost of living in various areas or provide special discounts to particular clientele, such as large purchasers or students.

Divide up your markets according to criteria like tastes, purchasing patterns, and income levels. Adjust pricing methods appropriately. For example, premium pricing can be used to target upscale consumers looking for premium cables, while penetration pricing can be used to rapidly penetrate new markets and increase market share. When it comes to pricing tactics, be adaptable and open to change in response to market trends and feedback.

HOW TO DETERMINE PROFIT MARGINS

To calculate your profit margin, deduct all of your costs (such as production, overhead, and marketing costs) from the money you make from selling cables. To find the profit margin percentage, divide the final profit by the revenue.

Maintaining a lucrative and sustainable firm requires maintaining a healthy profit margin. To make wise choices regarding price changes, cost reductions, and investment opportunities, keep a close eye on your profit margins. Maintaining good profit margins while remaining competitive requires keeping a watch on market changes and rivalry.

PROVIDING SALES AND MARKETING

Providing specials and discounts can be a useful tactic to draw clients and increase revenue. To encourage recurring purchases, take into account loyalty programs, bundle discounts, and seasonal specials.

Make sure that any discounts you offer don't materially affect your profit margins. Determine how discounts affect your overall sales and profitability. To generate excitement, get rid of inventory, or reward devoted consumers, use promotions wisely while keeping a healthy ratio of sales volume to profitability.

CHAPTER SIX

DISTRIBUTION AND SALES CHANNELS

INTERNET SALES: CHANNELS AND TECHNIQUES

Reaching clients across the globe is made possible by selling cables online. Select the best platform for your business to get started. E-commerce behemoths like Amazon, eBay, and Etsy are popular choices, as are building your website with tools like Shopify or Woo Commerce. Every platform has benefits, like Etsy's specialized market for handcrafted and one-of-a-kind goods or Amazon's enormous client base. To maximize your presence and sales, spend some time learning about the cost schedules, listing specifications, and marketing tools that each platform provides.

Digital marketing techniques must also be applied to sell products online effectively.

To increase search engine visibility, start by adding relevant keywords, thorough descriptions, and high-quality photos to your product listings. Promote your cables on social networking sites like Facebook, Instagram, and Twitter. Interact with potential consumers by posting frequently and offering promotions and interactive content. Building a subscriber list using email marketing enables you to deliver targeted promotions and information about new items, keeping your consumers informed and interested. Email marketing may also be a powerful tool.

Finally, think about using internet advertising to increase traffic to your listings. Pay-per-click (PPC) advertising on social media networks like Google Ads might draw in customers who are specifically looking for your kind of product. Make changes to your campaigns' performance using analytics tools to increase return on investment. Providing outstanding customer service will also help establish a good reputation and promote recurring business. This includes answering questions quickly and resolving returns and problems with efficiency.

OPPORTUNITIES FOR RETAIL AND WHOLESALE

Finding possible partners and learning about their needs is essential to growing your cable company into retail and wholesale markets. Start by looking at specialty shops, hardware stores, and local and national electronics stores that might sell your products. Make a persuasive case to buyers or shop managers that highlight the need, quality, and distinctiveness of your products. Providing volume discounts, competitive pricing, and eye-catching packaging will help draw retailers to your goods.

Opportunities that entail selling your cables in bulk to companies that will then distribute them to their clients might be especially profitable. Make a line sheet or wholesale catalog outlining your product options, costs, and conditions of sale to get started. To meet possible wholesale buyers, go to networking events and trade exhibits tailored to your sector. Professionalism, prompt communication, and the capacity to regularly meet wholesalers' volume and delivery requirements are necessary for developing relationships with them.

Furthermore, think about collaborating with online wholesale markets like Faire or Alibaba, which link suppliers with a variety of retail clients.

These sites can expedite the wholesale process and help you reach a worldwide audience. Make sure you can handle bulk orders and uphold superior standards to satisfy your wholesale partners and expand your business.

PARTICIPATING AT EXPOS AND TRADE SHOWS

Getting your cables seen at trade exhibits and expos is a great opportunity to meet people in the business, expand your network, and produce leads for sales. Start by picking the appropriate events, including hardware shows, specialized trade fairs, or consumer electronics expos, that correspond with your target market. To get a good booth placement, register early. You should also invest in a professional, eye-catching display that accentuates the qualities and advantages of your items. To draw attention to your exhibit, use banners, demos, and marketing supplies.

Making the most of trade show opportunities requires preparation. Bring lots of marketing brochures, business cards, and sample products. Prepare a succinct sales pitch and be prepared to interact with attendees by responding to their inquiries and providing comprehensive details about your services. To turn leads into sales, get interested attendees' contact information and swiftly follow up with them after the event. Providing discounts or show specials might encourage quick purchases and increase curiosity about your goods.

Making connections with other exhibitors and guests through networking can lead to new business prospects.

Engage in event-related activities, including panel discussions or seminars, to raise your profile and reputation in the business. Developing connections with influential people in the industry and possible partners can result in beneficial partnerships and recommendations, which can help you expand your business outside of trade shows.

CREATING ALLIANCES WITH DISTRIBUTORS

Creating alliances with distributors can help you reach a wider audience and improve the efficiency of your sales process. Distributors buy your cables in bulk and resell them to merchants or final customers, serving as middlemen. Investigate and find distributors who specialize in electronics or related products to form these partnerships. Make a strong business plan that emphasizes the benefits, demand, and profitability of your wires when you approach them.

Mutually beneficial terms and open communication are essential for successful relationships. Agreements defining prices, minimum order quantities, terms of payment, and delivery schedules should be negotiated. Giving distributors marketing support—like product training and promotional materials—will enable them to sell your goods more successfully. Keep lines of communication open to discuss possible new product lines or opportunities, track sales performance, and resolve any difficulties that may arise.

Strong, enduring partnerships with distributors are based on dependable supply chains, consistently high-quality products, and first-rate customer support. Ask your distributors for feedback on a proactive basis, and be willing to make changes in response to their suggestions. Review your agreements frequently to make sure they continue to be equitable and advantageous for all sides, building a partnership that propels your company's expansion and success.

ORGANIZING STOCK AND SHIPPING

To meet customer demand without overstocking or experiencing shortages, effective inventory management is essential. First things first, put in place an inventory management system that keeps tabs on sales, reorders, and stock levels. Automating these procedures with software programs like TradeGecko, Ordoro, or QuickBooks can give you access to real-time data and support you in making well-informed restocking decisions. Review your inventory frequently to spot trends and modify your ordering habits accordingly.

Another crucial component of any cable company is shipping. Select reputable shipping companies that provide reasonable prices and guarantee on-time arrival. Your inventory management and e-commerce systems can be integrated with platforms like Ship Station or Easy ship to streamline the fulfillment process. It is important to inform your clients about shipping policies, charges, and delivery periods to set realistic expectations. Various shipping choices, including expedited, international, and normal, can be provided to meet the needs of various clientele.

It's critical to package your cables properly to avoid damage during transit. To attract clients that care about the environment, invest in high-quality packing materials and take into account eco-friendly choices. To improve the customer experience, including tracking information and proper labeling with every order. In addition to increasing operational effectiveness, effective inventory and shipping management also promote customer satisfaction and repeat business.

CHAPTER SEVEN

PROMOTION AND MARKETING

FORMULATING A MARKETING STRATEGY

Setting goals is the first step in developing an effective marketing strategy for a newbie selling wires. With your cable business, what are your goals? Determine the needs of your target market and the places where they are most likely to search for cables. Investigate your rivals to find out what they have to offer and how you can set yourself apart.

Next, list your tactics in outline form. This covers marketing techniques, distribution routes, and price plans. Choose the positioning of your cables in the market and the features that will set them apart from the competition.

Set aside money for your marketing campaigns and distribute resources appropriately.

By carrying out your strategies, put your plan into action. This could entail running marketing campaigns, attending events or trade exhibits, and making use of web resources. To maximize your marketing efforts, keep a close eye on your progress and modify your plans in response to customer feedback and industry developments.

MAKING USE OF INTERNET AND SOCIAL MEDIA ADS

Social media is an effective marketing tool for cable companies. Produce interesting material that highlights your cables and speaks to the people who will be using them. Utilise social media sites like Twitter, Facebook, and

Instagram to share product images, videos, client endorsements, and sales. Interact with your audience by quickly answering messages and comments.

Invest in web advertisements to connect with more people. You may target particular interests and demographics with platforms like social media advertising and Google Ads. To encourage clicks and conversions, use attention-grabbing ad language and images. For best results, continuously assess the efficacy of your ads and modify your targeting and messaging.

To construct integrated campaigns, combine internet ads with social media content. For instance, to boost interaction and create leads, hold a social media giveaway or contest with promotional advertisements. To gauge the success of your efforts and make informed decisions, monitor KPIs like reach, engagement, clicks, and conversions.

BLOGGING AND CONTENT MARKETING

To draw in and keep your target audience interested, content marketing entails producing insightful and pertinent content. Create a blog on your website to provide helpful articles, tutorials, advice, and insights from the cable sector. Use pertinent keywords when optimizing your content for search engines to increase visibility and organic traffic.

To accommodate a variety of tastes, provide material in a variety of formats, including podcasts, infographics, films, and case studies. Spread the word about your work via email newsletters, industry forums, and social media to increase your reach and establish yourself as a reliable knowledge source in the cable sector.

Employ content marketing to highlight your experience, answer frequently asked questions, and inform your audience about the advantages of your services. To promote a sense of community and loyalty, ask readers to engage and provide comments. To improve your content strategy over time, track metrics related to content performance such as views, shares, and comments.

NEWSLETTERS AND EMAIL CAMPAIGNS

Email marketing campaigns are a powerful tool for nurturing leads, establishing trust, and introducing potential consumers to your cables. Create an email list of people who are interested in cables by using your website, social media accounts, and events. For more focused advertising, divide your list into segments according to demographics, hobbies, and past purchases.

Provide interesting and customized email content, such as special offers, discounts, product updates, and instructional materials. To draw readers in and promote opens, create attention-grabbing subject lines and images. Incorporate conspicuous calls-to-action (CTAs) that direct readers to your website or virtual store to complete a transaction.

Use email marketing solutions to automate your email campaigns so that you can send timely and pertinent communications based on subscriber behavior and preferences. Monitor important performance indicators like as open, click-through, conversion, and unsubscribe rates to assess the effectiveness of your campaigns and refine your email marketing approach for greater outcomes.

DEVELOPING CONNECTIONS AND NETWORKING

To establish connections and broaden your influence in the cable sector, networking is essential. To network with manufacturers, distributors, suppliers, and prospective clients, and attend trade exhibitions, conferences, and networking events. Establish rapport by having thoughtful discussions and sharing contact details.

To stay informed about market trends, opportunities, and difficulties, join associations for the cable sector, participate in forums, and visit online communities. Engage in conversations, impart your knowledge, and look for chances to work together with other industry individuals and companies.

To cross-promote your cables and reach new client segments, form alliances and collaborations with businesses that complement your own, such as technology companies, home improvement stores, and merchants of electronics. Develop enduring relationships with current clients by providing exceptional client care, offering loyalty plans, and sending out tailored communications to promote recommendations and repeat business.

CHAPTER EIGHT

CUSTOMER ASSISTANCE AND SUPPORT

OFFERING TOP-NOTCH CUSTOMER SUPPORT

In the cable manufacturing and distribution industry, timeliness and helpfulness are critical components of outstanding customer service. Make sure that your website and packaging have easy-to-find contact information, including phone numbers and email addresses. Answer consumer questions as soon as possible, and take the initiative to resolve any problems they might have with your products. Providing a variety of customer service avenues, such as live chat or a special help website, can also improve the general customer experience.

Customer satisfaction can also be greatly increased by teaching your customer support staff about typical problems and how to troubleshoot them. To address clients' concerns in a timely and efficient manner, always strive to show empathy and understanding in your contact with them. Offering consumers comprehensive and lucid product documentation, such as FAQs and troubleshooting manuals, can also enable them to resolve minor difficulties alone and minimize the need for intensive help.

Finally, think about putting in place a method for customers to rank and comment on their experiences. Make the required changes to your customer service procedures based on the input you've received. You may establish a solid reputation for providing exceptional customer service in the cable sector by continuously aiming to go above and beyond what customers anticipate and swiftly attending to their requirements.

REFUNDS AND COMPLAINTS ARE HANDLED

Refunds and complaints must be handled skillfully in the cable manufacturing and sales industry to keep customers happy and loyal. Clearly define your policies for refunds, swaps, and returns, and make sure your consumers are aware of them. As much as possible, make the return procedure easy for customers by offering options for swaps or refunds by their preferences, along with guidelines for returning products.

When handling complaints, use an empathic attitude to every circumstance and concentrate on coming up with a solution that pleases the client. Ask clarifying questions, actively listen to their issues, and offer your honest apologies for any trouble they may have had.

Collaborate with the client to determine the underlying cause of the problem and provide workable solutions that successfully meet their needs.

Keep meticulous records of all returns and grievances to monitor patterns and spot persistent problems that could need remediation. To lower returns and complaints over time, use this data to continuously improve your procedures and goods. You can use difficult circumstances to forge stronger bonds with customers and raise satisfaction levels by treating returns and complaints with professionalism and consideration.

CONSTRUCTING A LOYALTY SCHEME

In the highly competitive world of cable production and sales, creating a loyalty program can be an effective tactic to keep clients and promote repeat business. Establish the goals of your loyalty program first, like encouraging recurring purchases, rewarding recommendations, or providing special discounts and benefits to devoted clients. Create a tiered system that gives customers more perks as they move through the program, based on their level of engagement and purchases.

To encourage customer engagement, promote your loyalty program through a variety of platforms, including social media, email marketing, and in-store signage. Make sure to explain to clients the advantages of signing up for the program as well as how they can accrue and use rewards. To improve your loyalty program's value proposition, think about collaborating or partnering with companies that complement your own.

Engage members of your loyalty program regularly with tailored messages, unique events, and special offers. To learn about participants' preferences and make ongoing improvements to the program's offerings, ask for feedback from the participants. You may build enduring relationships in the cable industry that encourage advocacy and repeat business by encouraging a sense of community and rewarding client loyalty.

OBTAINING AND EXAMINING CLIENT INPUT

It's a good idea to collect and evaluate customer feedback if you want to improve the entire customer experience and your cable products. Use a variety of feedback channels, including social media listening, product reviews, and surveys, to get insights from a range of consumer demographics.

Offer rewards to consumers who submit feedback, or make the procedure simple and quick.

Sort and classify consumer reviews in an orderly manner to find recurring themes, emerging trends, and areas in need of development. To learn more about the preferences, problems, and expectations of your customers, use the qualitative data from open-ended responses. Measuring the impact of changes made in response to customer input and prioritizing areas of emphasis can be aided by quantitative data from ratings and analytics.

To get useful insights from vast amounts of consumer data, apply feedback analysis tools and strategies including sentiment analysis and trend monitoring. Work together across functional boundaries within your company to exchange input and insights, then create focused action plans for bettering services and products. In the cable sector, you can promote ongoing innovation and customer happiness by paying close attention to what customers have to say and acting upon your observations.

ENHANCING PRODUCTS IN RESPONSE TO CUSTOMER INPUT

A key element of success in the production and distribution of cables is the incorporation of client feedback into product improvement.

Begin by gathering input from a variety of sources, such as reviews, consumer surveys, and in-person contacts. Examine this feedback to find trends and recurrent recommendations for new features or improvements to the product that suit the needs and preferences of the users.

Sort feedback items into priority lists according to their possible impact and practicality for implementation. Create a product enhancement roadmap that details the precise modifications, the deadlines, and the accountable team members. Work collaboratively with the engineering, quality assurance, and product development teams to ensure high-quality results and quick execution of enhancements.

Inform clients clearly and openly about impending product changes, emphasizing how their input has shaped these advancements. Before making improvements widely available, ask a small number of consumers for input and beta testing. After a product is implemented, track customer satisfaction metrics and feedback trends to evaluate its efficacy and make any revisions.

Throughout the product's lifetime, solicit consumer feedback regularly to drive iterative changes and sustain market relevance. Strengthening client loyalty, increasing revenue, and setting yourself apart in the cutthroat cable market may all be achieved by using consumer data to improve your products iteratively.

CHAPTER NINE

GROWING YOUR COMPANY

GROWING YOUR PRODUCT OFFERING

To find new growth prospects, expanding your product range requires strategic planning and market research. To find out what products are in demand, start by examining consumer reviews and industry developments. To reach a wider audience, you can also look at complementing products that go well with your current line of business. For example, if you are selling USB cables, you could want to branch out into audio cables, HDMI cables, or adapters to better serve your clients' varied demands.

Conduct feasibility studies to evaluate each product's manufacturing viability, market demand, and profitability after you've identified possible new products. This includes assessing possible sales volumes, pricing schemes, and production costs. Work together with manufacturers and suppliers to make sure the transition to producing new items goes smoothly. Use focus groups and customer surveys to get their opinions on possible new items, then use their feedback to improve your current ones.

Using marketing campaigns, discounts, and cross-selling opportunities, carefully introduce new items to your current consumer base.

Track sales figures, get input, and refine features or designs of products in response to user feedback. You can efficiently extend into new market segments and bolster your brand presence by consistently broadening and varying your product line.

EMPLOYING AND EDUCATING PERSONNEL

Hiring and training employees is essential for sustaining operational effectiveness and producing high-quality products as your company expands. As a starting point, clearly define employment positions and duties according to your company's needs.

Create job descriptions that detail the education, training, and experience needed for each position. Employ social media, professional networks, and online job boards to draw in talented applicants who share the values and ethos of your business.

Establish a systematic hiring procedure that includes reference checks, interviews, and resume screening to evaluate applicants in-depth. After you've chosen qualified applicants, provide them with thorough onboarding training to ensure a smooth transition. This includes introducing new employees to the policies, practices, and product lines of your business. Provide employees with opportunities for continuous training and development to improve their abilities and output.

Establish a welcoming atmosphere at work that encourages cooperation, open communication, and teamwork. Employee feedback should be encouraged to pinpoint areas that need work and quickly resolve any issues rose.

By making hiring and training investments, you create a skilled and driven workforce that supports the expansion and success of your company.

CONTRACTING OUT PRODUCTION

There are many advantages to outsourcing production, such as scalability, cost savings, and access to specialized knowledge. Examine your production procedures and pinpoint areas where outsourcing can increase productivity and simplify operations. For instance, you could delegate cable assembly to a specialized manufacturer that employs knowledgeable specialists and cutting-edge machinery, freeing you up to concentrate on your primary business operations, including sales and marketing.

Make sure your outsourcing partners match your production schedules, budget, and quality standards by doing extensive due diligence before choosing them. Clear contracts that include deliverables, deadlines, cost breakdowns, and quality assurance procedures should be negotiated. Sustain consistent correspondence and observation to monitor advancement and swiftly resolve any concerns.

Beyond production, outsourcing can also involve distribution, warehousing, and logistics. If you want to save shipping costs and improve supply chain management, think about collaborating with outside logistics companies. You can increase operational effectiveness and concentrate on strategic growth goals by strategically outsourcing non-core functions.

PURCHASING NEW TECHNOLOGY

Keeping up with changing customer expectations and maintaining competitiveness needs investments in new technology. Determine whether technical innovations—such as automated production machinery, quality control systems, and inventory management software—are pertinent to your sector. Analyze the possible advantages and return on investment of implementing these technologies to boost output, cut expenses, and improve product quality.

Track performance indicators and solicit input to evaluate how technology expenditures affect your company's operations.

Keep an eye out for new developments in technology and industry trends to stay ahead of the curve and be flexible when the market shifts. Work together with technology providers and business leaders to investigate creative ideas that can boost productivity and creativity in all of your company's operations.

PLANNING AND MANAGEMENT OF FINANCES

Planning and managing finances well is essential to long-term profitability and growth. Make a thorough budget first, including estimates for income, costs, capital expenditures, and cash flow. Regularly track performance and pinpoint potential for cost-savings or improvement by keeping an eye on important financial measures.

To automate operations related to bookkeeping, invoicing, and financial reporting, make use of accounting systems or financial software. Establish strong internal controls to reduce risks and guarantee adherence to legal obligations. Create backup plans and risk control techniques to deal with unforeseen expenses or market volatility.

Examine your alternatives for funding capital expenditures and business expansion, including lines of credit, loans, and equity investments. Collaborate closely with consultants or financial advisors to maximize profits, minimize debt, and optimize tax methods. You may successfully scale your firm over the long run and handle economic swings by prioritizing prudent financial planning and management techniques.

CHAPTER TEN

FAQS & FREQUENTLY ASKED QUESTIONS

SOLVING ISSUES WITH MANUFACTURING

Production can be disrupted in the cable manufacturing industry by problems such as variable quality, shortages of materials, or equipment faults. Carrying out routine quality inspections during the production process is a good place to start when troubleshooting these issues. Ascertain that production equipment is calibrated and maintained properly and that raw materials fulfill specifications.

Inventory management needs to be proactive to address shortages of materials. Keep adequate stock levels by demand projections and cultivate a rapport with dependable suppliers. By putting preventative maintenance programs into place, machinery downtime caused by malfunctions is reduced. To maintain efficient production, train employees on troubleshooting typical problems and have backup plans ready for any setbacks.

HANDLING COMPETITION IN THE MARKET

Differentiating your wires is essential to standing out in a crowded market. To learn about the products, services, and pricing policies of rivals, conduct market research. Make use of this data to find areas where your product line needs to be improved or has gaps.

To draw clients, emphasize distinctive selling propositions like higher quality, cutting-edge features, or environmentally friendly materials. By highlighting your value proposition and strengths in marketing campaigns, you may establish a powerful brand identity. To win over customers' trust and loyalty, provide superior customer service and warranty policies.

TAKING CARE OF LEGAL AND COMPLIANCE CONCERNS

Manufacturing and selling cables requires adherence to industry standards and regulations. Learn about pertinent legislation, including those about environmental protection, safety, and labeling requirements. Make sure your items pass testing and certification requirements to stay out of trouble with the law and keep customers satisfied.

Make records of compliance, safety data sheets, and product specifications. Adopt quality control procedures to maintain safety regulations and product integrity.

Review and update your compliance procedures regularly to keep up with evolving laws and market trends.

RESPONDING TO INQUIRIES FROM CUSTOMERS

Having good customer communication is essential for answering questions and developing connections. Make sure that the product information on your website, packaging, and marketing materials is correct and clear. Make troubleshooting guides and FAQs to assist consumers in solving common problems on their own.

Provide a variety of customer service channels, including live chat, email, and phone, and people who are qualified to respond quickly to questions. Put in place a mechanism for collecting customer feedback so that you may use it to improve your offerings. To keep your reputation intact, respond to client complaints in a timely and professional manner.

GETTING READY FOR UPCOMING INNOVATIONS AND TRENDS

Keep abreast of market developments and consumer preferences to stay ahead of industry trends and innovations. Keep an eye on technical developments that might affect cable performance, design, or manufacturing procedures. Make research and development investments to create new products and enhance current ones.

Adjust to shifting consumer needs, such as the growing need for eco-friendly materials or the incorporation of intelligent technologies.

Examine potential alliances or joint ventures with tech firms to take advantage of skills and broaden your customer base.

www.ingramcontent.com/pod-product-compliance
Lightning Source LLC
Chambersburg PA
CBHW052331220526
45472CB00001B/372